The four rules of decimal money

by K A Hesse

Longman

Contents

Coin values and conversions

Addition

Subtraction

Multiplication

Division

£1 = 100 new pennies

Bronze coins

1 New Penny

a **one**

1p

½ New Penny

a **half**

½p

2 New Pence

a **two**

2p

Silver coins

5 New Pence

a **five**

10 New Pence

a **ten**

50 New Pence

a **fifty**

2

Say or write the values missing from these coins.

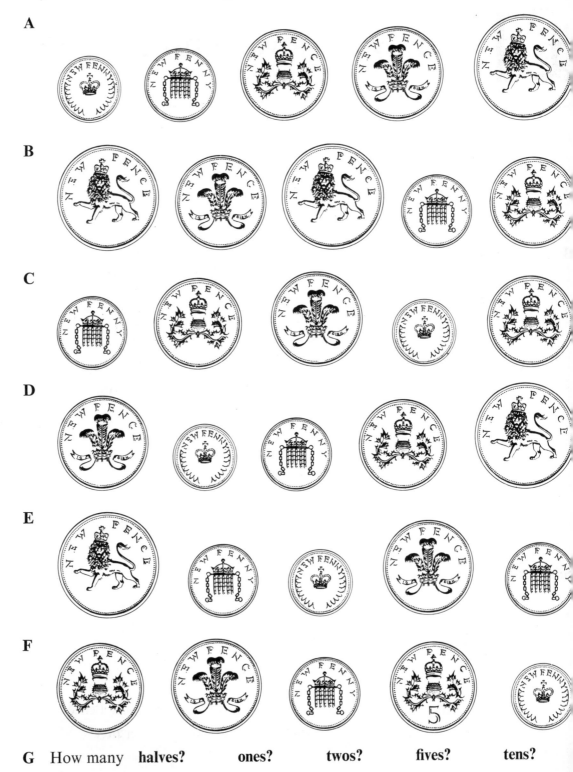

A

B

C

D

E

F

G How many **halves?** **ones?** **twos?** **fives?** **tens?**

For each line say or write first the value of the coin worth most and then the value of the coin worth least.

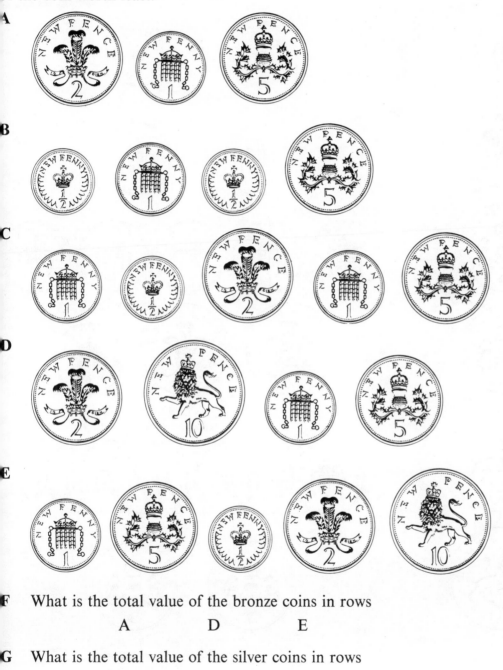

F What is the total value of the bronze coins in rows

        A   D   E

G What is the total value of the silver coins in rows

        B   C   E

4

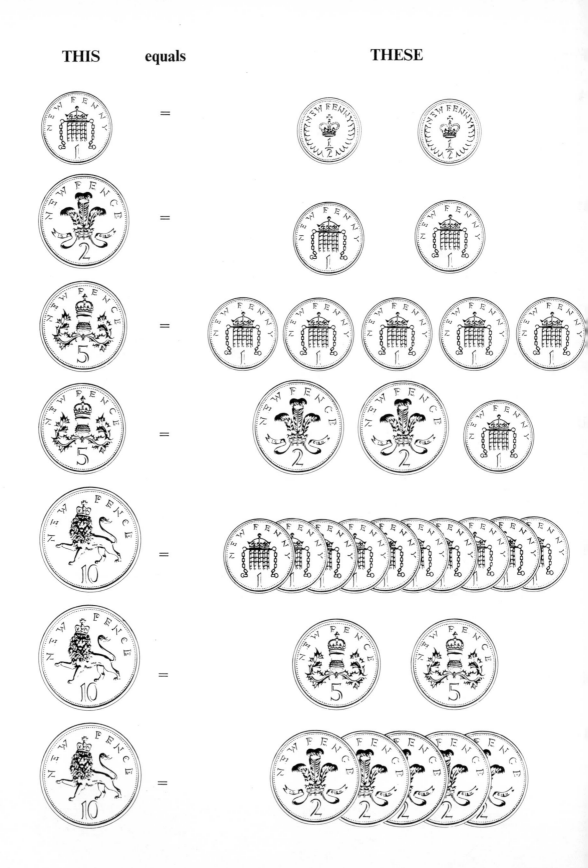

For each row find a single coin to equal the total value of the coins in it.

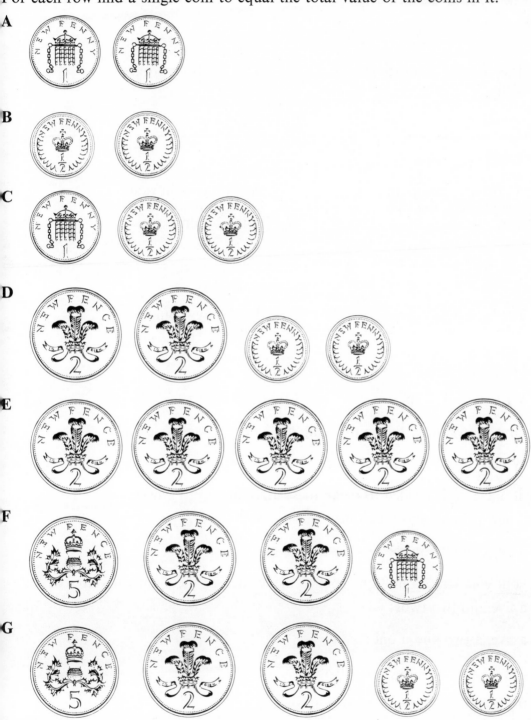

Draw these coins as well as you can, colour them and be sure to put the value or each.

A The largest bronze coin.

The smallest bronze coin.

B A bronze coin, in size between the two in A.

A silver coin in value equal to 10 of the last coin you drew.

C The smallest silver coin.

The only coin not round.

D Write in full: 2p   10p

E Give the value of the coin having only a crown as its design.

F Which coins have milled edges?

Which one coin is needed to make these amounts up to a five?

G 2 new pennies and a two new pence piece:

H 3 ones and 2 halves:

Use your coin tray to find one coin to make each of these amounts up to a ten.

I a five and two twos:

J a five, a two and a one:

K a five, two ones and two halves:

L three twos, two ones and two halves:

What are the missing coin values?

A

B (5p) = (p)(p)(p)   7p= (p)(p)(p)

(10p) = (2p)(p)(p)(p)

C 13p= (p)(p)(p)   20p= (p)(p)

32p= (p)(p)(p)(p)

D £3 = ___ fifties   8 fifties = ___

50p = 35p + (p)(p)   66p= (50p)(p)(p)(p)

Write as pounds:

E two pounds twenty-six _____   four pounds seventy _____

three pounds eight _____   eighty pence _____

18p _____   107p _____   300p _____

Write as spoken:

F £1·34 _____   £4·05 _____

£0·68 _____   £0·09 _____

A   On squared paper make a number square by writing out the numbers from 1 to 10 as the first row, and fill in row after row until you reach 100.

1	2	3	4	5	6	7			
11									
		23							

B   Count 2 and colour blue the square for number 2.

C   Count two more squares and colour number 4.

D   Go on counting in twos and colouring the squares up to 30.

E   Count in twos from 2 to 16, 8 to 40, 22 to 50.

What is the total value of each row of coins?

F

G

Find the missing numbers, counting in twos:

H   2, 4, 6, ___,___,___,___,___,___, 20.

I   14, 16, 18, ___,___,___,___,___,___,___,___,___,___.

J   Look at your square and count in twos from   1 to 21,   5 to 23, 15 to 40,   35 to 50.

Find the last number in each series, counting in twos:

K   5, 7,___,___,___, 15,___,___,___,___.

L   25, 27, ___,___,___,___,___,___,___,___,___,___,___.

Go back to your number square.

A   Count five and colour the square red.

B   Count five more squares and colour that square red.

You should have red squares for numbers 5 and 10.

C   Go on counting in fives and colouring the square red up to 50.

D   Count in fives from   5 to 30,   15 to 45,   35 to 70.

What is the total value of each of these rows of coins?

E

F

What are the missing numbers, counting in fives?

G   5, 10,____,____,____,____,____, 40.

H   20, 25,____,____,____,____,____,____,____, 65.

I   35, 40,____,____,____,____,____,____,____,____, 85.

J   75,____,____,____,____,____,____, 110.

Look at your square again.

K   Count in fives from   2 to 30,   8 to 33,   24 to 50.

Write the numbers in these series, counting in fives:

L   14,____,____,____,____,____,____, 49.

M   16,____,____,____,____,____,____, 51.

N   27,____,____,____,____,____,____,____,____, 72.

O   33,____,____,____,____,____,____,____,____, 78.

Again go back to your number square.

A   Count in tens from   10 to 50,   30 to 70,   50 to 100.

Find the value of each of these rows of coins:

B

C

Again go back to your number square.

D   Count in tens from   4 to 34,   7 to 57,   33 to 38.

E   28 to 78,   32 to 102,   45 to 125.

Find the total amount of the coin values in each row:

F

5p 5p 5p 10p 10p 10p 10p 10p

G

2p 2p 2p 2p 10p 10p 10p 10p 10p 10p

H

2p 2p 2p 5p 5p 5p 5p 5p 5p 5p

I

2p 2p 2p 2p 2p 2p 5p 5p 5p 10p

J

5p 5p 5p 5p 5p 2p 2p 2p 10p 10p

Find numbers missing from some rings by using your coin box.

A
(2p) = (1p) (p)   (1p) = (p) (p)

B
(2p) = (1p) (p) (p)   (5p) = (2p) (1p) (p)

C
(5p) = (2p) (1p) (p) (p)

D
(5p) = (2p) (2p) (p) (p)

E
(5p) = (2p) (1p) ($\frac{1}{2}$p) (p) (p)

F
(10p) = (5p) (2p) (p) (p)

G
(10p) = (2p) (2p) (p) (p)

H
(10p) = (5p) (1p) (1p) (p) (p)

I
(10p) = (2p) (2p) (2p) (p) (p)

J
(10p) = (2p) (1p) (p) (p)

K
(10p) = (5p) (2p) (2p) (p) (p)

You will need

2 fives,   3 twos   and   3 ones.

Use your coins to find which of them will be needed to pay each of these amounts exactly.
The rings show how many coins will be needed each time.
Work across the page.

A 3p = (p) (p)   4p = (2p) (p)

B 6p = (p) (p)   6p = (2p) (p) (p)

C 7p = (5p) (p) (p)   7p = (p) (p)

D 8p = (5p) (p) (p)   8p = (p) (p) (p) (p)

E 9p = (p) (p) (p)   9p = (p) (p) (p) (p)

F 10p = (p) (p)   10p = (p) (p) (p) (p)

G 11p = (p) (p) (p)   13p = (p) (p) (p) (p)

H 12p = (p) (p) (p) (p) (p)

I 15p = (p) (p) (p) (p) (p)

Tens, fives, twos, and ones

You will need: 4 tens   2 fives   3 twos   3 ones.

Use your coins to find which of them will be needed to pay each of these amounts.
Work across the page.

A

12p = (p) (p)   14p = (p) (p) (p) (p)

B

17p = (p) (p) (p)   17p = (p) (p) (p) (p)

C

20p = (p) (p)   18p = (p) (p) (p) (p)

D

22p = (p) (p) (p)   20p = (p) (p) (p)

E

25p = (p) (p) (p)   23p = (p) (p) (p) (1p)

F

30p = (p) (p) (p)   35p = (p) (p) (p) (p)

G

40p = (p) (p) (p) (p) (p)

H

27p = (p) (p) (p) (p) (p)

I

36p = (p) (p) (p) (p) (p) (p)

J

45p = (p) (p) (p) (p) (p)

This coin   equals   5 tens
                              10 fives
                              25 twos

A   How many fifties equal   100p?   £1?   £2?

B   Put a fifty on a balance.
        Does its weight balance 5 tens?   10 fives?

C   Put a ten on the balance.
        Does its weight balance 2 fives?   5 twos?

D   Put a two on the balance.
        Does its weight balance 2 ones?   4 halves?

The answers to **B**, **C** and **D** should make you curious.
Write a few sentences about your answers.

State the total values of these coins:

E

F

G

Copy this frame of squares:

5					30				50
50									

A   In the top row write the numbers from   5 to 50,   counting in fives.

B   In the second row write the numbers from   50 to 500, counting in fifties.

C   Write the full series, counting in fifties:

200,_____,_____,_____,_____,_____, 500 ,_____,_____,_____,_____.

How many pennies in one pound?

E   How many fifties in one pound?   £2?   £5?

F   How many pounds equal 3 fifties?   6 fifties?   20 fifties?

G   How many tens equal 1 fifty?   2 fifties?   10 fifties?

H   How many fives equal 1 fifty?   2 fifties?   6 fifties?

I   Find how many fifties there are, with tens over, in
7 tens   11 tens   17 tens

Find the missing values of coins needed to make up these amounts:

J   (50p) =40p+ (p)   (50p) =45p+ (p)

K   (50p) =30p+ (p)(p)   60p= (p)(p)

L   (50p) =20p+ (p)(p)(p)

M   (50p) =70p− (p)(p)

N   (50p) =65p− (p)(p)   £1= (p)(p)

* *

A comma was used at one time to separate thousands from hundreds.   Now we
use a space.   i.e. 2 476.
Rewrite these numbers, showing the space:

A   4760   7030   60790

One hundred new pence = one pound.

In amounts of 100p and over we need to show the pounds separate from the pence.

B   What do we use to show where one sentence ends and another begins?

We do the same to show where pounds in a number end and the pence begin.

For 132p we write £1·32.
In 132 pence we have £1 for the 100p and there are 32p left over.
This means that we put the £ sign before the 1 and the full stop after it.

So   132p   becomes   £1·32

Note: We do not put in the p when we use the £ sign and we place the full stop half
        way up the figure and not on the line.
        We do not call it a full stop in that position but a POINT.

We WRITE   £1·32

We SAY   One pound thirty-two

Put in the pound sign and the point to each of these amounts of pence:

C   1 4 2   2 4 6   3 0 7   5 8 6   1 9 2   3 5 7   1 0 9

Put in the pound sign and the point.   Then write the amount in words.

D   1 3 5   2 1 6

E   4 7 2   8 9 5

For   206p   you would have   £2·06 (two pounds six)
For   260p   it would be   £2·60 (two pounds sixty)

The place-holder 0 in £2·60 makes sure that we do not mistake it for two pounds
and six pence.

Write these amounts in figures:

F   One pound twenty-nine   One pound forty-five

G   One pound seventy   Two pounds eight

H   Three pounds four   Five pounds ten

I   One pound one   Six pounds seventeen

These numbers increase by ten at a time.   Put in the missing numbers.

A   85,____,____,____,____,____ , 145, 155

B   73,____,____,____,____,____ , 133, 143

C   90,____,____,____,____,____,____ , 160

These amounts increase by ten pence a time.   Put in the missing amounts:

D   80p, 90p, £1,____,____,____,____ , £1·50

E   64p, 74p, 84p,____,____,____,____,____ , £1·44

F   77p, 87p,____,____,____,____,____,____ , £1·57

Write the three amounts which come after these, increasing by 1p each time.

G   £1·07   £2·04

H   £2·09   £3·18

I   98p   £1·08

J   £1   £2

Write in words:

K   £2·35   £4·70

L   £1·08   £1·03

M   £3·10   £2·09

Increase each of these amounts by four pence:

N   £1·18   £1·16   97p   96p   £1·98   £10

If we take £1 from £1·25 we are left with £·25.
Someone taking a quick look at £·25 might mistake it for £25.
To help against this we write
     £0·25   always using a 0 before the point when there are no pounds figures.
We would even write
     £0·06 for six pence.

Write these amounts as pounds:

O   32p   70p   16p   8p   60p

P   15p   4p   40p   107p   7p

Q   10p   208p   20p   120p   102p

Pence to pounds

To change 200p to pounds we write only £2.

Write as pounds:

A	300p	500p	660p	404p	90p
B	60p	600p	800p	6p	200p

Copy these columns and put in what is missing:

	As pence	As pounds	As spoken
C	427p	£4·27	four pounds twenty-seven
D		£0·50	fifty pence
E	4730p		forty-seven pounds thirty
F	9p		ninepence
G	103p		
H	40p		
I		£0·80	
J	800p		eight pounds
K		£15·07	
L	1610p		
M			seventeen pence
N			one pound five and a half pence
O			forty pounds seventeen
P		£20·10	
Q	708p		
R		£0·04$\frac{1}{2}$	
S			nineteen pence
T	1900p		
U			twenty-eight and a half pence

				Further practice

A 19p + 3p =   4p + 5p + 3p =   17p + 4p + 6p = | *page 20*

B Beneath each answer rewrite it in true pound form.

12p	35p	27p	36p
20p	19p	8p	68p
57p	46p	75p	97p
——	——	——	——

page 21

C $1\frac{1}{2}$p + 1p = $\frac{1}{2}$p + $\frac{1}{2}$p + $\frac{1}{2}$p = $3\frac{1}{2}$p + $1\frac{1}{2}$p = | *pages 22 & 23*

D Write down in columns and add, stating all answers in pounds.

36p + 26p + 43p $24\frac{1}{2}$p + $7\frac{1}{2}$p + $56\frac{1}{2}$p

$47\frac{1}{2}$p + 30p + $56\frac{1}{2}$p   78p + $50\frac{1}{2}$p + 49p | *page 24 A/F*

E Add:

£2·56	£3·78	£1·68$\frac{1}{2}$	£0·67$\frac{1}{2}$
0·08	0·60$\frac{1}{2}$	0·08$\frac{1}{2}$	3·09
1·70	0·09$\frac{1}{2}$	0·23$\frac{1}{2}$	0·28$\frac{1}{2}$
——	——	——	——

page 24 G/J

F Write down in columns and add:

£1·53 + £0·69 + £2·08   £15·76 + 40p + £3·84

five pounds fifteen + thirty-six pence + forty-nine pence | *page 25 A/F*

G Write down in columns and add:

Seven pounds sixty-one plus thirty-three pence plus
one pound sixteen plus two pounds.

One pound eight plus five pounds plus two pounds ninety-three.

Twenty-three pounds sixty-six, plus one pound eighty and a half
plus ninety-three and a half, plus ten pounds. | *page 25 J/N*

Find the value of these groups of coins:

Add: work across the page.

D   a five + a two + a half =   a five + a one + a half =

E   a five + a two + a two =   a ten + a five + a one =

F   a ten + a ten + a five =   a five + a two + a ten =

G   a ten + a two + a five =   a ten + a one + a five =

H   3p + 5p =   4p + 2p =   8p + 1p =   6p + 2p =
                                4p + 5p + 1p =

I   7p + 2p =   9p + 2p =   6p + 5p =   10p + 2p =
                                6p + 2p + 2p =

J   8p + 2p =   11p + 2p =   11p + 5p =   8p + 5p =
                                9p + 2p + 5p =

K   9p + 5p =   10p + 5p =   7p + 5p =   12p + 2p =
                                11p + 1p + 5p =

L   12p + 5p =   13p + 2p =   17p + 2p =   13p + 5p =
                                14p + 5p + 2p =

M   14p + 2p =   16p + 5p =   19p + 2p =   18p + 5p =
                                17p + 2p + 5p =

Add: work across the page.

A	7p + 5p =	17p + 5p =	27p + 5p =	37p + 5p + 2p =
B	6p + 4p =	16p + 4p =	36p + 4p =	56p + 4p + 5p =
C	8p + 5p =	18p + 5p =	28p + 5p =	48p + 5p + 5p =
D	11p + 5p =	21p + 5p =	51p + 5p =	71p + 5p + 10p =
E	17p + 5p =	27p + 5p =	47p + 5p =	67p + 5p + 10p =

Which of these amounts are below one pound?

F   49p   52p   83p   101p   203p   97p   107p

Write each of these amounts as pence:

G   £1·25   £1·30   £0·85   £1·06   £0·07   £0·10

Write each of these amounts as pounds:

H   146p   170p   26p   30p   83p   90p

I   207p   60p   101p   8p   10p   7p

Add each of these sums to get an answer in pence.
Beneath each answer rewrite it in true pound form.

Work across the page.

J	21p +93p	32p +85p	24p +97p	47p +76p	35p +57p	47p +68p
	———	———	———	———	———	———

K	40p +73p	36p +80p	27p +90p	40p +80p	36p +74p	58p +62p
	———	———	———	———	———	———

L	63p 20p 59p	23p 42p 36p	26p 6p 76p	36p 77p 89p	37p 65p 99p	96p 9p 98p
	———	———	———	———	———	———

A Which of these can be cut into halves?

 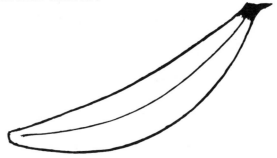 

B How many halves in one apple?   2 apples?

C How many halves in one banana?   3 bananas?

We have a halfpenny coin, so there is no need to cut any pennies.

D How many halfpennies are worth 1 penny?

E How many halfpennies are worth 2 pennies?   3 pennies?

F How many halfpennies are equal to these coins?

G How many pennies are worth these groups of coins?

H How many pence are worth these groups of coins?

Count the total value of these coins:

A

B

How many halves equal these amounts?

C	2p	4p	7p	$2\frac{1}{2}$p	$3\frac{1}{2}$p	$5\frac{1}{2}$p
D	3p	5p	$6\frac{1}{2}$p	8p	$9\frac{1}{2}$p	12p

How many pence equal these amounts?

E	3 halves	4 halves	5 halves	9 halves
F	$2 \times \frac{1}{2}$p =	$6 \times \frac{1}{2}$p =	$7 \times \frac{1}{2}$p =	$10 \times \frac{1}{2}$p =
G	8 halves	11 halves	16 halves	20 halves
H	12 halves	15 halves	13 halves	21 halves

Work out these sums: work across the page.

I	$\frac{1}{2}$p + $\frac{1}{2}$p =	$1\frac{1}{2}$p + $\frac{1}{2}$p =	$2\frac{1}{2}$p + $\frac{1}{2}$p =	$1\frac{1}{2}$p + $1\frac{1}{2}$p =
J	$2\frac{1}{2}$p + 1p =	$3\frac{1}{2}$p + 2p =	$3\frac{1}{2}$p + $\frac{1}{2}$p =	$2\frac{1}{2}$p + $1\frac{1}{2}$p =
K	$\frac{1}{2}$p + 5p =	$\frac{1}{2}$p + $1\frac{1}{2}$p =	$\frac{1}{2}$p + $6\frac{1}{2}$p =	$1\frac{1}{2}$p + $3\frac{1}{2}$p =
L	$\frac{1}{2}$p + $\frac{1}{2}$p + $\frac{1}{2}$p =	$1\frac{1}{2}$p + $\frac{1}{2}$p + $\frac{1}{2}$p =	2p + $1\frac{1}{2}$p + $1\frac{1}{2}$p =
M	$\frac{1}{2}$p + $1\frac{1}{2}$p + $\frac{1}{2}$p =	2p + $1\frac{1}{2}$p + $2\frac{1}{2}$p =	$1\frac{1}{2}$p + $4\frac{1}{2}$p + $\frac{1}{2}$p =
N	$2\frac{1}{2}$p + $\frac{1}{2}$p + $3\frac{1}{2}$p =	$4\frac{1}{2}$p + 5p + $1\frac{1}{2}$p =	$3\frac{1}{2}$p + $1\frac{1}{2}$p + $2\frac{1}{2}$p =
O	$1\frac{1}{2}$p + $2\frac{1}{2}$p + $1\frac{1}{2}$p =	$3\frac{1}{2}$p + 2p + $2\frac{1}{2}$p =	4p + $3\frac{1}{2}$p + $2\frac{1}{2}$p =

Add each of these sums, then rewrite the answer in pounds.
Work across the page:

A

$43\frac{1}{2}$p	$37\frac{1}{2}$p	$16\frac{1}{2}$p	$49\frac{1}{2}$p	$35\frac{1}{2}$p	$54\frac{1}{2}$p
$+54\frac{1}{2}$p	$+60\frac{1}{2}$p	$+74\frac{1}{2}$p	$+60$ p	$+65$ p	$+47$ p

B

$7\frac{1}{2}$p	$20\frac{1}{2}$p	$34\frac{1}{2}$p	$40\frac{1}{2}$p	$52\frac{1}{2}$p	$75\frac{1}{2}$p
$32\frac{1}{2}$p	14 p	$7\frac{1}{2}$p	$9\frac{1}{2}$p	$40\frac{1}{2}$p	9 p
$21\frac{1}{2}$p	$38\frac{1}{2}$p	78 p	$65\frac{1}{2}$p	$7\frac{1}{2}0$	$15\frac{1}{2}$p

Write down in columns and add, stating all answers in pounds.

C $47p + 23p + 16p$ $\qquad$ $35p + 17p + 54p$ $\qquad$ $65p + 70p + 68p$

D $25\frac{1}{2}p + 64\frac{1}{2}p + 30p$ $\qquad$ $30\frac{1}{2}p + 8\frac{1}{2}p + 64p$ $\qquad$ $34\frac{1}{2}p + 9\frac{1}{2}p + 56\frac{1}{2}p$

E $34p + 68\frac{1}{2}p + 97\frac{1}{2}p$ $\qquad$ $23p + 80p + 97p$ $\qquad$ $45\frac{1}{2}p + 9\frac{1}{2}p + 45p$

F $66\frac{1}{2}p + 25p + 9\frac{1}{2}p$ $\qquad$ $37p + 50p + 19p$ $\qquad$ $68p + 70\frac{1}{2}p + 65p$

Add: work across the page.

G

172	287	306	275	306	193
+276	+404	+278	+126	+298	+207

H

£1·72	£2·87	£3·06	£2·75	£3·06	£1·93
+ 2·76	+ 4·04	+ 2·78	+ 1·26	+ 2·98	+ 2·07

I

£2·30	£1·06	£3·46	£2·60	£3·80	£2·86
+ 1·73	+ 2·94	+ 0·70	+ 0·47	+ 0·76	+ 3·14

J

£0·26$\frac{1}{2}$	£1·05$\frac{1}{2}$	£0·34	£1·67	£0·08$\frac{1}{2}$	£1·68$\frac{1}{2}$
1·31$\frac{1}{2}$	1·30	1·06$\frac{1}{2}$	0·08$\frac{1}{2}$	2·86$\frac{1}{2}$	0·97
0·23$\frac{1}{2}$	0·64$\frac{1}{2}$	0·70	1·30	0·09$\frac{1}{2}$	0·34$\frac{1}{2}$

Add (+): work across the page.

A	£1·29	£2·47	£1·80	£0·72	£2·93
	2·96	0·34	0·75	2·08	3·07
	1·30	4·52	0·35	3·20	2·80

B	£2·32	£13·40	£5·07	£14·60	£25·83
	4·06	5·08	0·85	3·08	0·79
	10·35	0·57	12·98	20·32	3·38

Write down in columns and add:

C £1·36 + £0·72 + £2·04   £0·48 + £1·50 + £2·17

D £1·24 + 76p + £2·20   £2·83 + £7·08 + 69p

E £2·80 + 52p + 70p   £1·80 + 36p + 84p

F £1·12 + 80p + 8p   £0·73 + 90p + 7p

G thirty-six and a half pence + eight pence + sixty-six and a half pence.

H two pounds twenty-six + seventy pence + one pound four.

I ten pounds eighty-five + six pounds nine + three pounds six.

Add: work across the page.

J	£1·47	£2·46	£3·85	£0·67	£27·86
	3	1·54	2	5·08	16
	1·33	3	0·15	0·25	0·19

Write down in columns and add:

K £2·70 + 43p + £1·08 + £3   £1·05 + 86p + £2 + £1·09

L £4·76 + £2 + £10·18 + £3·06   48p + £10·87 + £3 + 65p

M one pound eight + eighty-seven pence + three pounds five + five pounds.

N twenty-six pounds fifteen + seventy-eight pence + nine pounds seven + four pounds.

Check your subtraction

Further practice

A $2p - \frac{1}{2}p =$ $3\frac{1}{2}p - 2\frac{1}{2}p =$ $5p - 1\frac{1}{2}p =$ $6p - 3\frac{1}{2}p =$ *page 27*

B Under each sum rewrite the answer in its true pound form:

£3·67	£4·03	£1·02	£2·60
− 1·05	− 3·86	− 0·38½	− 2·09½

page 28

C

£5·30½	£15·21	£2·03½	£4·11
− 2·08½	− 14·93	− 0·97	− 1·08½

page 29

D

£3·18	£5·20½	£12·03	£10·03
− 1·07½	− 0·96½	− 11·92½	− 0·93

page 30

E

£1·03	£4·00	£7·00	£20·00
− 0·08	− 2·73	− 0·91	− 19·96

page 31 A/D

F Write in columns and subtract:

Take two pounds four from five pounds twelve.

From seven pounds take four pounds eight.

Ten pounds minus three pounds ninety-six.

Subtract nineteen pounds ninety-one from thirty pounds.

page 31 E/P

How many halves would equal each of these coins?

A

B What would be the change after paying $\frac{1}{2}$p for a toffee monster from coins having these values?

    $\frac{1}{2}$p   2p   1p   5p   10p

C What would be the change from each of these coins after paying $1\frac{1}{2}$p for a bar of toffee?

         2p   5p   10p

D What would be left from each of these amounts after paying $1\frac{1}{2}$p for a bar of toffee?

Work out these sums:

E $1p - \frac{1}{2}p =$ $1\frac{1}{2}p - \frac{1}{2}p =$ $1\frac{1}{2}p - 1p =$ $\frac{1}{2}p - \frac{1}{2}p =$
     $2p - 1p =$

F $2p - \frac{1}{2}p =$ $1\frac{1}{2}p - 1\frac{1}{2}p =$ $2p - 1\frac{1}{2}p =$ $2\frac{1}{2}p - \frac{1}{2}p =$
     $3\frac{1}{2}p - \frac{1}{2}p =$

G $2\frac{1}{2}p - 1p =$ $2\frac{1}{2}p - 1\frac{1}{2}p =$ $3\frac{1}{2}p - 1\frac{1}{2}p =$ $3p - 1\frac{1}{2}p =$
     $3\frac{1}{2}p - 2p =$

H $3\frac{1}{2}p - 2\frac{1}{2}p =$ $4p - \frac{1}{2}p =$ $4p - 3\frac{1}{2}p =$ $4p - 1\frac{1}{2}p =$
     $4p - 2\frac{1}{2}p =$

I $4\frac{1}{2}p - 2p =$ $4\frac{1}{2}p - 1\frac{1}{2}p =$ $5p - 4\frac{1}{2}p =$ $5p - 1\frac{1}{2}p =$
     $5p - 3\frac{1}{2}p =$

J $3\frac{1}{2}p - 3\frac{1}{2}p =$ $5p - \frac{1}{2}p =$ $5p - 2\frac{1}{2}p =$ $4p - 2\frac{1}{2}p =$
     $6p - 4\frac{1}{2}p =$

Look at £0·56.

When we write a number of pence less than 100p as pounds we must remember to
put a place-holder 0 before the decimal point, as

£0·62 for "sixty-two pence".
(Remember £1·62 is said as, "One pound sixty-two".)

£·62 may be carelessly read as £62 instead of 62p and
£·17 may be carelessly read as £17 instead of 17p, so £0·62 and £0·17 prevent that

Write as pence:

A	£2·42	£0·69	£0·78	£10·34
B	£0·27	£10·58	£20·85	£0·96

Write as pounds:

C one pound fourteen   twenty-three pence

D eighty-seven pence   ninety-two pence

E four pounds sixteen   fifty-one pence

Take away:

Under each sum re-write the answer in its true pound form:

F	£2·63 − 1·51	£3·81 − 3·63	£4·02 − 3·46	£7·53 − 6·78	£5·64 − 4·69
G	£4·32 − 2·35	£15·43 − 10·67	£11·27 − 10·65	£8·08 − 7·39	£10·26 − 9·78
H	£7·04 − 6·75	£10·05 − 9·27	£11·06 − 10·08	£13·02 − 12·09	£3·67 − 2·88
I	£0·87½ − 0·35	£0·63½ − 0·15½	£3·58 − 2·23½	£2·06 − 1·34½	£1·40 − 1·07½

Look at £2·05.

When we write a number of pence less than 10p as pounds or with pounds we must put a place-holder 0 after the decimal point to show that it is a numeral in the units column, as

£1·06 for one pound and six pence (said, "One pound six".)
or £0·08 for eight pence.

Write as pence:

A £1·08 £0·18 £1·04 £0·05 £0·09

Write as pounds:

B one pound two sixpence

C eighty-four pence sevenpence

D two pounds three elevenpence

E eighteen pence one pound seven

Take away:

Then under each sum re-write the answer in its true pound form:

F
£3·27	£12·42	£4·21	£10·63	£5·07
− 1·25	− 8·37	− 3·17	− 9·68	− 4·03
———	———	———	———	———

G
£7·06	£10·04	£11·06	£8·05	£10·08
− 5·50	− 4·98	− 8·98	− 7·90	− 9·90
———	———	———	———	———

H
£6·00	£11·00	£12·65	£10·33	£5·38
− 3·92	− 9·07	− 0·97	− 8·29	− 0·30
———	———	———	———	———

I
£2·32½	£2·70½	£3·11	£1·34	£3·08
− 2·03	− 1·64½	− 1·06½	− 0·20½	− 0·70½
———	———	———	———	———

Look at £2·70.

When the figures to the right of the decimal point show one of the group of tens, as 10, 20, 30, etc., we must put in the place-holder 0 to show that the numeral is in the tens column and not the units column, as

£1·40 for one pound and forty pence (said, "One pound forty".)

£1·4 could be said easily as, "One pound four", so we write £1·40 for one pound forty and £1·04 for one pound four.

Write as pence:

A £1·60 £1·06 £0·16 £10·16 £0·80

Write as pounds:

B two pounds twenty two pounds twenty   one pound fifty

C three pounds thirty seventy pence

D four pounds ten sixty pence

Take away:

Under each answer rewrite it in its true pound form:

E	£2·35	£4·62	£3·48	£5·06	£8·05
	− 1·05	− 0·72	− 0·78	− 0·56	− 7·15
	────	────	────	────	────

F	£10·57	£10·06	£2·30	£10·08	£11·70
	− 3·07	− 9·60	− 1·80	− 0·88	− 0·99
	────	────	────	────	────

G	£8·40	£11·32	£10·04	£12·06	£7·08
	− 7·90	− 8·02	− 0·94	− 0·98	− 6·98
	────	────	────	────	────

H	£1·30	£10·63	£14·05	£11·78½	£10·08½
	− 0·08½	− 1·52½	− 0·98½	− 10·90	− 9·98½
	────	────	────	────	────

Subtract: work across the page.

A	£7·16 – 4·07	£9·06 – 6·36	£5·50 – 0·83	£7·20 – 1·08	£6·60 – 2·07

B	£3·05 – 1·95	£5·08 – 0·90	£2·04 – 0·96	£1·06 – 0·99	£3·97 – 0·89

C	£6·01 – 4·08	£4·03 – 2·07	£2·03 – 0·09	£3·06 – 0·08	£1·07 – 0·09

D	£3·00 – 1·63	£5 – 2·76	£4 – 1·06	£6 – 1·93	£10 – 6·92

Write in columns and subtract:

E £5·46 – £3·40   £2·70 – £1·06   £3·02 – £1·08

F £7·30 – £6·25   £10 – £4·84   £7 – £3·96

G Three pounds minus one pound fifteen.

H Four pounds eight minus two pounds thirty-eight.

I Two pounds thirty-three minus seventy-seven pence.

J From ten pounds take six pounds nineteen.

K Deduct three pounds ninety-eight from five pounds twenty-four.

L Deduct ninety-seven pence from thirteen pounds forty.

M Deduct two pounds ninety-three from eleven pounds.

N From twenty-three pence subtract fourteen and a half pence.

O Subtract one pound six and a half from three pounds twelve.

P Deduct one pound eight and a half from four pounds.

Check your multiplication

Further practice

A Work out each sum then rewrite it in its true pound form:

$$\begin{array}{r} 24\text{p} \\ \times\quad 7 \\ \hline \end{array} \qquad \begin{array}{r} 70\text{p} \\ \times\quad 8 \\ \hline \end{array} \qquad \begin{array}{r} 96\text{p} \\ \times\quad 11 \\ \hline \end{array} \qquad \begin{array}{r} 59\text{p} \\ \times\quad 12 \\ \hline \end{array} \qquad \begin{array}{r} 90\text{p} \\ \times\quad 12 \\ \hline \end{array}$$

page 33

B Multiply each of these by 10 and then by 100:

4p   20p   80p

page 34

C £1·23 × 10 =   £0·09 × 100 =

£0·04½ × 10 =   £0·20½ × 100 =

page 35

D $$\begin{array}{r} 4\frac{1}{2}\text{p} \\ \times\quad 5 \\ \hline \end{array} \qquad \begin{array}{r} 3\frac{1}{2}\text{p} \\ \times\quad 9 \\ \hline \end{array} \qquad \begin{array}{r} 5\frac{1}{2}\text{p} \\ \times\quad 8 \\ \hline \end{array} \qquad \begin{array}{r} 9\frac{1}{2}\text{p} \\ \times\quad 7 \\ \hline \end{array} \qquad \begin{array}{r} 11\frac{1}{2}\text{p} \\ \times\quad 8 \\ \hline \end{array}$$

page 36
A/I

E Under each answer rewrite it in its true pound form:

$$\begin{array}{r} 14\frac{1}{2}\text{p} \\ \times\quad 9 \\ \hline \end{array} \qquad \begin{array}{r} 27\frac{1}{2}\text{p} \\ \times\quad 8 \\ \hline \end{array} \qquad \begin{array}{r} 45\frac{1}{2}\text{p} \\ \times\quad 11 \\ \hline \end{array} \qquad \begin{array}{r} 67\frac{1}{2}\text{p} \\ \times\quad 12 \\ \hline \end{array}$$

page 36
J/L

F What is five times one pound six?

Multiply one pound sixty-two and a half by eight.

What amount is equal to twelve times one pound eight and a half?

page 37

G What is the total of seven times thirty-five and a half pence?

What amount in pounds equals nine times eight and a half pence?

Is eleven times one pound eighty and a half greater or less than twenty pounds?

page 38

Write answers as pounds.   Work across the page:

A	7p × 4	6p × 5	4p × 8	5p × 7
B	6p × 8	8p × 7	5p × 9	7p × 12
C	8p × 8	9p × 10	10p × 8	8p × 9
D	7p × 10	11p × 9	8p × 12	10p × 10
E	6p × 12	9p × 9	7p × 9	12p × 8
F	9p × 8	10p × 9	8p × 11	10p × 11
G	9p × 11	11p × 10	12p × 9	9p × 12
H	10p × 12	12p × 12	11p × 11	12p × 11

Under each answer rewrite it in its true pound form.
Work across the page:

I

$$\begin{array}{r} 17p \\ \times\ 5 \\ \hline \end{array} \qquad \begin{array}{r} 14p \\ \times\ 7 \\ \hline \end{array} \qquad \begin{array}{r} 27p \\ \times\ 9 \\ \hline \end{array} \qquad \begin{array}{r} 19p \\ \times\ 6 \\ \hline \end{array} \qquad \begin{array}{r} 17p \\ \times\ 6 \\ \hline \end{array} \qquad \begin{array}{r} 15p \\ \times\ 7 \\ \hline \end{array}$$

J

$$\begin{array}{r} 24p \\ \times\ 5 \\ \hline \end{array} \qquad \begin{array}{r} 35p \\ \times\ 8 \\ \hline \end{array} \qquad \begin{array}{r} 20p \\ \times\ 9 \\ \hline \end{array} \qquad \begin{array}{r} 40p \\ \times\ 10 \\ \hline \end{array} \qquad \begin{array}{r} 60p \\ \times\ 7 \\ \hline \end{array} \qquad \begin{array}{r} 50p \\ \times\ 6 \\ \hline \end{array}$$

K

$$\begin{array}{r} 50p \\ \times\ 8 \\ \hline \end{array} \qquad \begin{array}{r} 60p \\ \times\ 5 \\ \hline \end{array} \qquad \begin{array}{r} 80p \\ \times\ 5 \\ \hline \end{array} \qquad \begin{array}{r} 67p \\ \times\ 9 \\ \hline \end{array} \qquad \begin{array}{r} 88p \\ \times\ 8 \\ \hline \end{array} \qquad \begin{array}{r} 86p \\ \times\ 7 \\ \hline \end{array}$$

L

$$\begin{array}{r} 75p \\ \times\ 8 \\ \hline \end{array} \qquad \begin{array}{r} 87p \\ \times\ 7 \\ \hline \end{array} \qquad \begin{array}{r} 67p \\ \times\ 12 \\ \hline \end{array} \qquad \begin{array}{r} 89p \\ \times\ 9 \\ \hline \end{array} \qquad \begin{array}{r} 75p \\ \times\ 12 \\ \hline \end{array} \qquad \begin{array}{r} 96p \\ \times\ 11 \\ \hline \end{array}$$

M

$$\begin{array}{r} 69p \\ \times\ 9 \\ \hline \end{array} \qquad \begin{array}{r} 95p \\ \times\ 11 \\ \hline \end{array} \qquad \begin{array}{r} 68p \\ \times\ 8 \\ \hline \end{array} \qquad \begin{array}{r} 95p \\ \times\ 12 \\ \hline \end{array} \qquad \begin{array}{r} 78p \\ \times\ 9 \\ \hline \end{array} \qquad \begin{array}{r} 85p \\ \times\ 12 \\ \hline \end{array}$$

4 by itself is worth 4 units.   Move it one place to the left and put in a place-holder 0 and we have 40, in which the 4 is worth four tens.   Move it another place to the left and use another place-holder 0 and we have 400, in which the 4 is worth four hundreds.   If we move 56 one place to the left we have 560—ten times bigger, and moving 72 two places to the left we get 7 200—one hundred times bigger.

A   Write or say what the 8 is worth in each of these numbers:

3 826

5 287

408

We find the 8 is being moved to the right and becoming ten times less each time. Finish these rules:

B   "To move a number one place to the left makes it _____ "

C   "To move a number one place to the right makes it _____ "

D   "To move a number two places to the left makes it _____ "

E   "To move a number two places to the right makes it _____ "

Use your rules to multiply these numbers by 10:

F   23p   6p   30p   204p   70p   50p

Use your rules to multiply these numbers by 100:

G   7p   9p   41p   20p   8p   60p

Let us use the rules to multiply £1·57 by 10.   When moved one place to the left we get

£15·70

by moving the figure after the point to the place before it and bringing in the place-holder 0 to show that there are 70p and not 7p.
Multiplying £0·63 by 10 we get £6·30, remembering that the 0 before the point is no longer needed when there is a figure to take its place, but we must bring in the place-holder 0 after the 3 to make clear that it represents 30p and not 3p.

Multiply each of these amounts by 10 and by 100:

H   £2   £7   8p   2p

I   £5   3p   7p   £9

J   6p   £0·07   £0·01   £0·08

K   5p   £0·03   9p   £0·10

Multiply each of these amounts by 10 and by 100:

A	£0·47	£0·83	£0·46
B	£1·50	£7·60	£3·08
C	£5·60	£0·09	£0·04
D	£0·30	£2·10	£0·06
E	£0·75	£2·06	£0·80
F	£1·09	£0·07	£2·03

Multiply: work across the page:

G $\frac{1}{2}p \times 2 =$ $\frac{1}{2}p \times 4 =$ $\frac{1}{2}p \times 3 =$ $\frac{1}{2}p \times 5 =$

H $\frac{1}{2}p \times 6 =$ $\frac{1}{2}p \times 7 =$ $\frac{1}{2}p \times 9 =$ $\frac{1}{2}p \times 8 =$

I $\frac{1}{2}p \times 10 =$ $\frac{1}{2}p \times 100 =$

J $2p \times 10 =$ $\frac{1}{2}p \times 10 =$ $2\frac{1}{2}p \times 10 =$ $2\frac{1}{2}p \times 100 =$

K $4p \times 10 =$ $\frac{1}{2}p \times 10 =$ $4\frac{1}{2}p \times 10 =$ $4\frac{1}{2}p \times 100 =$

L $3p \times 10 =$ $3\frac{1}{2}p \times 10 =$ $\frac{1}{2}p \times 100 =$ $3\frac{1}{2}p \times 100 =$

M $7p \times 10 =$ $7\frac{1}{2}p \times 10 =$ $\frac{1}{2}p \times 100 =$ $7\frac{1}{2}p \times 100 =$

N $12p \times 10 =$ $12\frac{1}{2}p \times 10 =$ $12\frac{1}{2}p \times 100 =$ $14\frac{1}{2}p \times 100 =$

O $15p \times 10 =$ $15\frac{1}{2}p \times 10 =$ $15p \times 100 =$ $15\frac{1}{2}p \times 100 =$

In the following sums write answers first in pence and then in pounds:

P $\frac{1}{2}p \times 10 =$ $\frac{1}{2}p \times 100 =$ $1\frac{1}{2}p \times 100 =$

Q $1\frac{1}{2}p \times 10 =$ $2\frac{1}{2}p \times 10 =$ $2\frac{1}{2}p \times 100 =$

R $4\frac{1}{2}p \times 100 =$ $7\frac{1}{2}p \times 100 =$ $11\frac{1}{2}p \times 10 =$

S $11\frac{1}{2}p \times 100 =$ $16\frac{1}{2}p \times 100 =$ $10\frac{1}{2}p \times 10 =$

Some of the answers in P–S will help in the following:

T $£0·02\frac{1}{2} \times 10 =$ $£0·02\frac{1}{2} \times 100 =$ $£0·03\frac{1}{2} \times 10 =$

U $£0·03\frac{1}{2} \times 100 =$ $£0·06\frac{1}{2} \times 10 =$ $£0·07\frac{1}{2} \times 100 =$

V $£1·04\frac{1}{2} \times 10 =$ $£1·08\frac{1}{2} \times 100 =$ $£0·16\frac{1}{2} \times 100 =$

W $£0·12\frac{1}{2} \times 10 =$ $£0·15\frac{1}{2} \times 100 =$ $£0·11\frac{1}{2} \times 100 =$

X $£1·26\frac{1}{2} \times 10 =$ $£2·57\frac{1}{2} \times 100 =$ $£2·40\frac{1}{2} \times 100 =$

Multiply: work across the page.

A $\frac{1}{2}$p $\times$ 2 = $\frac{1}{2}$p $\times$ 3 = $\frac{1}{2}$p $\times$ 4 = $\frac{1}{2}$p $\times$ 6 =

B $\frac{1}{2}$p $\times$ 5 = $1\frac{1}{2}$p $\times$ 5 = $1\frac{1}{2}$p $\times$ 6 = $1\frac{1}{2}$p $\times$ 4 =

C $1\frac{1}{2}$p $\times$ 7 = $\frac{1}{2}$p $\times$ 8 = $1\frac{1}{2}$p $\times$ 8 = $2\frac{1}{2}$p $\times$ 8 =

D $2\frac{1}{2}$p $\times$ 3 = $3\frac{1}{2}$p $\times$ 3 = $2\frac{1}{2}$p $\times$ 4 = $1\frac{1}{2}$p $\times$ 7 =

E $3\frac{1}{2}$p $\times$ 5 = $1\frac{1}{2}$p $\times$ 11 = $1\frac{1}{2}$p $\times$ 10 = $4\frac{1}{2}$p $\times$ 6 =

F $2\frac{1}{2}$p $\times$ 6 = $5\frac{1}{2}$p $\times$ 12 = $4\frac{1}{2}$p $\times$ 7 = $2\frac{1}{2}$p $\times$ 9 =

G $1\frac{1}{2}$p $\times$ 12 = $5\frac{1}{2}$p $\times$ 11 = $2\frac{1}{2}$p $\times$ 12 = $7\frac{1}{2}$p $\times$ 4 =

H $\frac{1}{2}$p $\times$ 9 = $\frac{1}{2}$p $\times$ 7 = $\frac{1}{2}$p $\times$ 11 = $1\frac{1}{2}$p $\times$ 9 =

I $3\frac{1}{2}$p $\times$ 6 = $5\frac{1}{2}$p $\times$ 7 = $6\frac{1}{2}$p $\times$ 8 = $4\frac{1}{2}$p $\times$ 12 =

J
| $3\frac{1}{2}$p $\times$ 7 | $2\frac{1}{2}$p $\times$ 5 | $3\frac{1}{2}$p $\times$ 6 | $5\frac{1}{2}$p $\times$ 8 | $4\frac{1}{2}$p $\times$ 9 | $7\frac{1}{2}$p $\times$ 8 |

K
| $6\frac{1}{2}$p $\times$ 9 | $8\frac{1}{2}$p $\times$ 7 | $7\frac{1}{2}$p $\times$ 9 | $9\frac{1}{2}$p $\times$ 9 | $7\frac{1}{2}$p $\times$ 11 | $8\frac{1}{2}$p $\times$ 12 |

Under each answer re-write it in its true pound form.

L
| $13\frac{1}{2}$p $\times$ 7 | $17\frac{1}{2}$p $\times$ 8 | $15\frac{1}{2}$p $\times$ 9 | $21\frac{1}{2}$p $\times$ 6 | $33\frac{1}{2}$p $\times$ 8 | $18\frac{1}{2}$p $\times$ 5 |

M
| $20\frac{1}{2}$p $\times$ 6 | $30\frac{1}{2}$p $\times$ 9 | $26\frac{1}{2}$p $\times$ 11 | $33\frac{1}{2}$p $\times$ 9 | $40\frac{1}{2}$p $\times$ 8 | $28\frac{1}{2}$p $\times$ 7 |

N
| $17\frac{1}{2}$p $\times$ 12 | $27\frac{1}{2}$p $\times$ 8 | $24\frac{1}{2}$p $\times$ 9 | $25\frac{1}{2}$p $\times$ 11 | $19\frac{1}{2}$p $\times$ 7 | $37\frac{1}{2}$p $\times$ 12 |

Multiply: work across the page.

A

$$£1·23 × 3 \qquad £1·13 × 4 \qquad £1·23 × 5 \qquad £2·16 × 7 \qquad £2·21 × 6$$

B

$$£1·25\tfrac{1}{2} × 5 \qquad £2·16\tfrac{1}{2} × 4 \qquad £4·17\tfrac{1}{2} × 8 \qquad £3·56\tfrac{1}{2} × 7 \qquad £2·75\tfrac{1}{2} × 9$$

C

$$£3·40 × 6 \qquad £2·60 × 7 \qquad £1·50 × 8 \qquad £4·50 × 6 \qquad £5·76 × 9$$

D

$$£1·20\tfrac{1}{2} × 6 \qquad £2·70\tfrac{1}{2} × 8 \qquad £1·80\tfrac{1}{2} × 5 \qquad £2·86\tfrac{1}{2} × 9 \qquad £1·96\tfrac{1}{2} × 11$$

E

$$£2·07 × 7 \qquad £3·05 × 8 \qquad £2·05 × 12 \qquad £5·07 × 9 \qquad £4·03 × 11$$

F

$$£4·08\tfrac{1}{2} × 9 \qquad £2·19\tfrac{1}{2} × 12 \qquad £3·93\tfrac{1}{2} × 11 \qquad £2·09\tfrac{1}{2} × 12 \qquad £1·09\tfrac{1}{2} × 11$$

G Find what equals six times two pounds thirty-four and a half.

H Find what equals eight times three pounds seventy-five and a half.

I Multiply eleven pounds eighty by seven.

J Multiply eight pounds seventy-nine by nine.

K What amount equals eight times seven pounds fifty and a half?

L What amount equals eleven times ninety-eight and a half pence?

M Find if seven times two pounds fifteen is greater or less than fifteen pounds.

N Can six pounds eight and a half be subtracted nine times from fifty-five pounds?

Multiply: work across the page.

A £0·65 £0·48 £0·25 £0·38 £0·75
× 7 × 9 × 8 × 11 × 12

B £0·76½ £0·38½ £0·29½ £0·19½ £0·38
× 8 × 7 × 11 × 12 × 1

C £0·04 £0·07 £0·05 £0·08 £0·09
× 8 × 11 × 8 × 12 × 12

D £0·08½ £0·08½ £0·07½ £0·09½ £0·08
× 7 × 9 × 11 × 11 × 1

E £13·07 £11·07 £13·90 £10·97 £12·09
× 9 × 12 × 9 × 11 × 12

F £60·90 £3·07½ £75·86 £4·70½ £5·98
× 12 × 11 × 8 × 12 × 1

G What sum equals seven times three pounds fourteen?

H What is the total of six times five pounds nine and a half?

I Multiply eighty-four and a half pence by eight.

J What amount equals eleven times eighteen and a half pence?

K What is the product of seven pounds eighteen and twelve?

L Find the product, in pounds, of nine and a half pence and twelve.

M Multiply twenty and a half pence by eleven.

				Further practice
A	$7\overline{)80}$ $\qquad$ $11\overline{)103}$ $\qquad$ $9\overline{)160}$ $\qquad$ $12\overline{)1100}$			*page 40*
B	£0·78 ÷ 10 $\qquad\qquad\qquad$ £10·90 ÷ 100 £0·65 ÷ 10 $\qquad\qquad\qquad$ £0·80 ÷ 100			*page 41*
C	$5\overline{)£6·25}$ $\qquad$ $7\overline{)£10·01}$ $\qquad$ $9\overline{)£14·04}$ $\qquad$ $12\overline{)£23·04}$			*page 42*
D	$8\overline{)£2·56}$ $\qquad$ $9\overline{)£5·04}$ $\qquad$ $12\overline{)£0·96}$ $\qquad$ $11\overline{)£7·70}$			*page 43 A/E*
E	$7\overline{)£5·62}$ $\qquad$ $11\overline{)£19·82}$ $\qquad$ $9\overline{)£6·70}$ $\qquad$ $12\overline{)£7·22}$			*page 43 F/L*
F	$8\overline{)£6·03}$ $\qquad$ $9\overline{)£3}$ $\qquad$ $12\overline{)£7}$ $\qquad$ $11\overline{)£80}$			*page 44 A/D*
G	$10\tfrac{1}{2}$p ÷ 3 = $\qquad$ $58\tfrac{1}{2}$p ÷ 9 = $\qquad$ 78p ÷ 12 =			*page 44 E/J*
H	$18\tfrac{1}{2}$p ÷ 5 = $\qquad$ 70p ÷ 8 = $\qquad$ 85p ÷ 11 =			*page 45 A/G*
I	$4\overline{)£12·22}$ $\qquad$ $9\overline{)£0·50}$ $\qquad$ $11\overline{)£0·85\tfrac{1}{2}}$ $\qquad$ $12\overline{)£10·90}$			*page 45 H/M*
J	$\tfrac{1}{2}$ yard of ribbon at $28\tfrac{1}{2}$p per yard = $\tfrac{1}{4}$lb. of ham at 49p per pound $\quad$ = $1\tfrac{1}{2}$doz. eggs at $27\tfrac{1}{2}$p per dozen $\quad$ =			*page 46*

Division : preparation

Supply the missing numbers: work across the page.

A	$7 \times \underline{\quad} = 21$	$2 \times \underline{\quad} = 16$	$7 \times \underline{\quad} = 35$	$6 \times \underline{\quad} = 2$
B	$5 \times \underline{\quad} = 40$	$6 \times \underline{\quad} = 42$	$7 \times \underline{\quad} = 49$	$9 \times \underline{\quad} = 5$
C	$8 \times \underline{\quad} = 56$	$10 \times \underline{\quad} = 70$	$9 \times \underline{\quad} = 63$	$8 \times \underline{\quad} = 6$
D	$11 \times \underline{\quad} = 66$	$12 \times \underline{\quad} = 60$	$8 \times \underline{\quad} = 72$	$11 \times \underline{\quad} = 9$
E	$6 \times \underline{\quad} = 72$	$9 \times \underline{\quad} = 81$	$12 \times \underline{\quad} = 84$	$10 \times \underline{\quad} = 12$
F	$12 \times \underline{\quad} = 96$	$11 \times \underline{\quad} = 132$	$9 \times \underline{\quad} = 108$	$12 \times \underline{\quad} = 10$

Write answers only: work across the page.

G	$24 \div 4 =$	$26 \div 4 =$	$28 \div 7 =$	$31 \div 7 =$
H	$32 \div 8 =$	$37 \div 8 =$	$40 \div 9 =$	$41 \div 6 =$
I	$51 \div 6 =$	$49 \div 9 =$	$52 \div 8 =$	$50 \div 7 =$
J	$60 \div 7 =$	$60 \div 8 =$	$64 \div 9 =$	$70 \div 8 =$
K	$81 \div 10 =$	$81 \div 7 =$	$60 \div 12 =$	$70 \div 12 =$
L	$75 \div 8 =$	$80 \div 11 =$	$76 \div 9 =$	$96 \div 10 =$
M	$82 \div 9 =$	$98 \div 8 =$	$90 \div 12 =$	$103 \div 11 =$
N	$104 \div 8 =$	$110 \div 9 =$	$103 \div 12 =$	$111 \div 12 =$

Divide: work across the page.

O	$9\overline{)940}$	$7\overline{)803}$	$9\overline{)810}$	$6\overline{)900}$	$8\overline{)700}$
P	$7\overline{)606}$	$9\overline{)790}$	$8\overline{)850}$	$11\overline{)890}$	$12\overline{)905}$
Q	$8\overline{)7206}$	$12\overline{)1800}$	$9\overline{)5222}$	$12\overline{)5407}$	$11\overline{)4700}$
R	$9\overline{)8170}$	$12\overline{)1030}$	$9\overline{)9073}$	$11\overline{)1080}$	$12\overline{)1195}$

We have worked with rules about changing the value of a number by moving its figures to the right or to the left.

A   What is the value of the figure 3 in each of these numbers?

       £1 283   £4 379   £5 036

B   What is the value of the figure 6 in each of these amounts?

       £16·42   £1·64   £0·16

C   What is the value of the figure 7 in each of these amounts?

       £24·70   £247   £2·47

Complete:

D   To INCREASE the value of a number TEN times
                                 is the same as to _____ it by 10.

E   To DECREASE the value of a number TEN times
                                 is the same as to _____ it by 10.

F   To DECREASE the value of a number ONE HUNDRED times
                                 is the same as to _____ it by 100.

G   To DIVIDE a number by 10 we move the position of its figures one
                                 place to the _____ .

H   To DIVIDE a number by 100 we move the position of its figures two
                                 places to the _____ .

Write what each of these amounts becomes when divided by 10 and by 100:

I   £10   £20   £5   £2   £500

Write what each of these amounts becomes when divided by 10 or by 100, but remember we only have 2 figures after the point. Therefore the rest of the figures will be a number of pence left over.

       e.g.   £2·15 ÷ 10 = £0·21 r5p   £2·15 ÷ 100 = £0·02 r15p

J   £2·63 ÷ 10 =   £12·06 ÷ 10 =   £4·50 ÷ 10 =

K   £32·15 ÷ 100   £4·08 ÷ 100 =   £10·40 ÷ 100 =

L   £23 ÷ 10 =   £107 ÷ 100 =   £1·75 ÷ 100 =

M   £0·67 ÷ 10 =   £0·83 ÷ 100 =   £0·68 ÷ 100 =

Dividing pounds and pence

A How many tens are there in £1?   £3?   £7?

B How many pence are there in 1 ten?   4 tens?   9 tens?

C Divide 435p by 3 and write your answer in pounds.

Look at these diagrams:

```
1 hundred = 10 tens            1 pound = 10 tens
    1 ten = 10 units              1 ten = 10 ones
  H   T   U                     £   tens   ones
  4   3   5                     4   3       5
```

Whether we divide hundreds of pence or pounds and pence we can treat them both alike, for the relation of figures in one column to those in the other is the same in both groups.   In any number the place value of a figure is ten times the place value of the figure on its right.   When dividing pounds the figure left over can be carried in front of the next figure because each pound equals 10 tens.

We must remember to put the DECIMAL POINT in the ANSWER above its position in the DIVIDEND.

$$\begin{array}{r} £1 \cdot 45 \\ \overline{3)£4 \cdot 35} \end{array} \quad \text{is worked like} \quad \begin{array}{r} 145p \\ \overline{3)435p} \end{array} \quad £1 \cdot 45 = 145p$$

DIVIDE: work across the page.

D $2)\overline{£3 \cdot 12}$ $3)\overline{£7 \cdot 11}$ $4)\overline{£10 \cdot 64}$ $6)\overline{£15 \cdot 18}$ $5)\overline{£17 \cdot 40}$

E $7)\overline{£15 \cdot 05}$ $8)\overline{£10 \cdot 80}$ $7)\overline{£10 \cdot 08}$ $9)\overline{£20 \cdot 07}$ $8)\overline{£20 \cdot 08}$

F $9)\overline{£9 \cdot 27}$ $11)\overline{£11 \cdot 44}$ $12)\overline{£13 \cdot 08}$ $9)\overline{£9 \cdot 63}$ $12)\overline{£25 \cdot 08}$

G $8)\overline{£16 \cdot 56}$ $11)\overline{£12 \cdot 10}$ $9)\overline{£11 \cdot 70}$ $11)\overline{£16 \cdot 50}$ $12)\overline{£22 \cdot 80}$

H $7)\overline{£25 \cdot 20}$ $9)\overline{£18 \cdot 72}$ $8)\overline{£40 \cdot 16}$ $12)\overline{£31 \cdot 08}$ $11)\overline{£30 \cdot 03}$

Divide: work across the page.

A 7)£10·50   9)£6·66   8)£5·36   7)£7·28   9)£7·02

B 8)£8·72   7)£5·04   9)£0·63   11)£6·60   12)£10·80

C 9)£0·81   12)£11·40   11)£10·78   9)£7·92   12)£2·04

D 8)£0·72   9)£8·01   12)£0·96   11)£9·90   12)£1·08

E 9)£8·10   7)£6·30   11)£21·01   12)£13·08   12)£20·40

In these sums remember that we have two figures only after the point so you will have remainders.   Write them as pence.

Work across the page:

F   £2·41 ÷ 10 =   £3·75 ÷ 100 =   £2·60 ÷ 10 =

G   £3·80 ÷ 10 =   £1·80$\frac{1}{2}$ ÷ 100 =   £0·78 ÷ 10 =

H   £0·62 ÷ 10 =   £0·73$\frac{1}{2}$ ÷ 100 =   £5·08$\frac{1}{2}$ ÷ 10 =

I   £10·07 ÷ 10 =   £1·09$\frac{1}{2}$ ÷ 100 =   £6·04 ÷ 100 =

J 7)£3·62   9)£10·73   8)£11·06   7)£6·05   12)£8·09

K 8)£12·03   7)£3·04   9)£6·34   8)£20·03   11)£7·75

L 9)£12·10   11)£9·06   7)£5·61   9)£11·72   12)£19·23

Divide: work across the page.

A $12\overline{)£10\cdot09}$ $7\overline{)£0\cdot66}$ $9\overline{)£0\cdot83}$ $11\overline{)£1\cdot04}$ $12\overline{)£1\cdot10}$

B $7\overline{)£30}$ $9\overline{)£20}$ $8\overline{)£30}$ $12\overline{)£10}$ $11\overline{)£5}$

C $9\overline{)£4}$ $11\overline{)£10}$ $9\overline{)£8}$ $12\overline{)£60}$ $12\overline{)£9}$

D $7\overline{)£5}$ $12\overline{)£70}$ $9\overline{)£30}$ $12\overline{)£30}$ $11\overline{)£60}$

Dividing with pence and halfpence

E   Share 9p amongst 4 children.

Your answer should be $2\frac{1}{4}$p each   or   2p r1p.
We have no coin for $\frac{1}{4}$p.   Our smallest coin is worth $\frac{1}{2}$p.
We have no $\frac{1}{4}$p, so that does not help.

The 1p over can be changed into 2 halfpence but we still cannot share amongst 4 children, so we are left with 1p that cannot be shared.
Let us share   14p   amongst   4   children.

$$14p \div 4 = 3p \ r2p$$
BUT $2p = 4$ halfpence
     4 halfpence $\div 4 = 1$ halfpence each child
SO $14p \div 4 = 3\frac{1}{2}p$ for each child.

Do these sums by changing left-over pence into halfpence and dividing again to get the final answer.

F $12p \div 3 =$ $13\frac{1}{2}p \div 3 =$ $16p \div 4 =$ $18p \div 4 =$

G $15p \div 6 =$ $12\frac{1}{2}p \div 5 =$ $24\frac{1}{2}p \div 7 =$ $28p \div 8 =$

H $45p \div 10 =$ $40\frac{1}{2}p \div 9 =$ $45\frac{1}{2}p \div 7 =$ $60p \div 8 =$

I $37\frac{1}{2}p \div 5 =$ $60\frac{1}{2}p \div 11 =$ $78p \div 12 =$ $85\frac{1}{2}p \div 9 =$

J $68p \div 8 =$ $67\frac{1}{2}p \div 9 =$ $93\frac{1}{2}p \div 11 =$ $90p \div 12 =$

n these sums, after changing left-over pence into halfpence and dividing again, state what is finally left over as pence.

A	$23p \div 4 =$	$27\frac{1}{2}p \div 4 =$	$29p \div 5 =$
B	$25p \div 7 =$	$40p \div 6 =$	$45p \div 8 =$
C	$37p \div 10 =$	$51p \div 9 =$	$63p \div 11 =$
D	$65p \div 7 =$	$70p \div 12 =$	$70p \div 9 =$
E	$74p \div 8 =$	$74p \div 11 =$	$84p \div 10 =$
F	$85p \div 9 =$	$96p \div 10 =$	$90p \div 12 =$
G	$87p \div 7 =$	$99p \div 12 =$	$95p \div 11 =$

H	$4)\overline{£8{\cdot}50}$	$5)\overline{£6{\cdot}23}$	$6)\overline{£8{\cdot}07}$	$4)\overline{£10{\cdot}22}$	$8)\overline{£11{\cdot}31}$
I	$7)\overline{£9{\cdot}40}$	$6)\overline{£8{\cdot}14}$	$8)\overline{£2{\cdot}68}$	$9)\overline{£5{\cdot}05}$	$7)\overline{£4{\cdot}08}$
J	$6)\overline{£6{\cdot}34}$	$7)\overline{£7{\cdot}46}$	$6)\overline{£4{\cdot}06}$	$8)\overline{£9{\cdot}64}$	$9)\overline{£10{\cdot}86}$
K	$6)\overline{£7{\cdot}23}$	$9)\overline{£5{\cdot}81}$	$7)\overline{£11}$	$11)\overline{£6}$	$12)\overline{£7}$
L	$7)\overline{£1{\cdot}50\frac{1}{2}}$	$9)\overline{£1{\cdot}03\frac{1}{2}}$	$8)\overline{£3{\cdot}80\frac{1}{2}}$	$12)\overline{£2{\cdot}46}$	$11)\overline{£3{\cdot}35\frac{1}{2}}$
M	$9)\overline{£0{\cdot}77\frac{1}{2}}$	$12)\overline{£0{\cdot}80\frac{1}{2}}$	$9)\overline{£0{\cdot}94\frac{1}{2}}$	$11)\overline{£0{\cdot}86\frac{1}{2}}$	$12)\overline{£0{\cdot}92}$
N	$11)\overline{£100}$	$9)\overline{£71}$	$12)\overline{£5}$	$12)\overline{£25}$	$11)\overline{£23}$
O	$8)\overline{£8{\cdot}60}$	$12)\overline{£36{\cdot}66}$	$11)\overline{£8{\cdot}08}$	$9)\overline{£4}$	$12)\overline{£49{\cdot}50}$

Think about this:

Find the price of $\frac{1}{2}$kg of butter at $36\frac{1}{2}$p per kilo.

Here is a problem for the shopkeeper.

$$\tfrac{1}{2} \text{ of } 36\tfrac{1}{2}\text{p} = 18\text{p r}\tfrac{1}{2}\text{p}$$

If the shopkeeper charged 18p per $\frac{1}{2}$kg of butter some people would buy 2 ha kilos of butter instead of 1kg and so save $\frac{1}{2}$p each time.
So the charge for $\frac{1}{2}$kg of butter at $36\frac{1}{2}$p per kg will be $18\frac{1}{2}$p.
This takes the price to the next $\frac{1}{2}$p.

Find the shop price for these:

A $\frac{1}{2}$kg of pears at $16\frac{1}{2}$p per kilo. $\frac{1}{2}$litre of milk at $8\frac{1}{2}$p per litre.

B $\frac{1}{2}$kg of tomatoes at $28\frac{1}{2}$p per kg.   half a melon at $20\frac{1}{2}$p each.

C $\frac{1}{2}$doz. eggs at $26\frac{1}{2}$p per dozen. $\frac{1}{4}$kg of tea at 89p per kilo.

The shopkeeper has another brand of butter at $37\frac{1}{2}$p per kilo.

$$\tfrac{1}{2} \text{ of } 37\tfrac{1}{2}\text{p} = 18\tfrac{1}{2}\text{p r}\tfrac{1}{2}\text{p}$$

As there is a remainder of more than the $\frac{1}{2}$p in the answer the shopkeeper wi charge 19p per $\frac{1}{2}$kg, that is, he will take the price to the next penny.
Remember (a) if the left-over pence are not enough for $\frac{1}{2}$p the shopkeeper wi charge $\frac{1}{2}$p, but (b) if the left-over pence are enough for $\frac{1}{2}$p in the answer and ther is still some left over then the shopkeeper charges to the next penny.

Find the shop prices for these:

D $\frac{1}{2}$kg of apples at $19\frac{1}{2}$p per kg. $\frac{1}{2}$kg of bacon at $44\frac{1}{2}$p per kilo.

E $1\frac{1}{2}$kg of cabbage at $7\frac{1}{2}$p per kg. $\frac{1}{2}$kg of cherries at $21\frac{1}{2}$p per kg.

F $\frac{1}{4}$kg of raspberries at 57p per kg. $\frac{1}{4}$kg of prawns at 99p per kilo.

G $1\frac{1}{2}$kg of peas at $8\frac{1}{2}$p per kilo. $1\frac{1}{4}$kg of plaice at 73p per kilo.

H $\frac{3}{4}$kg of cheese at 45p per kg. $1\frac{1}{4}$kg of beans at $36\frac{1}{2}$p per kg.

I $\frac{1}{4}$kg of ham at $104\frac{1}{2}$p per kg. $\frac{3}{4}$kg of sausage at $33\frac{1}{2}$p per kg.